MW00845371

MAGIC OF THE SEA

Also by Max Albert Wyss
MAGIC OF THE WOODS

MAX ALBERT WYSS

MAGIC OF THE SEA

A STUDIO BOOK · THE VIKING PRESS · NEW YORK

Published in 1971 by The Viking Press, Inc.
625 Madison Avenue, New York, N.Y. 10022

Published in Canada by
The Macmillan Company of Canada Limited

SBN 670-44992-x

Library of Congress catalog card number:
75-139482

Printed and bound in Switzerland

Layout and design: Edmund Amstad

BEHOLD THE SEA,
THE OPALINE, THE PLENTIFUL
AND STRONG,
YET BEAUTIFUL AS IS THE
ROSE IN JUNE,
FRESH AS THE TRICKLING
RAINBOW OF JULY;
SEA FULL OF FOOD,
THE NOURISHER OF KINDS,
PURGER OF EARTH, AND
MEDICINE OF MEN;
CREATING A SWEET CLIMATE
BY MY BREATH,
WASHING OUT HARMS AND
GRIEFS FROM MEMORY,
AND, IN MY MATHEMATIC
EBB AND FLOW,
GIVING A HINT OF THAT
WHICH CHANGES NOT.

RALPH WALDO EMERSON

MAGIC OF THE SEA

There is an age-old urge in man that drives him to trace the courses of great rivers past weirs and waterfalls and over bridges, to follow the unceasing flow of moving water on and on until, at last, he reaches the coast and—finally—the sea.

Every coast is something of a final destination, an insuperable boundary, or so at least it seems to those whose roots are fixed and settled in populated areas of land. Faced for the first time with a huge expanse of sea, the land-dweller may feel bewildered and perplexed by so much emptiness, the breath of which is redolent with loneliness and desolation, like high plains swept bare by the wind. And the ground crumbles underfoot....

Cold green whirlpools lurk menacingly beneath the refuge of overhanging reefs, and the seething water eats like acid at the rocks. Fountains of spray, in the clarity of slow motion, show where the breakers strike, and the pale sand on the rocks bears fragments of shells and green-brown seaweed in the deceptive stillness of the ebbing tide.

He who is new to the sea may experience a new and unknown rapture; or the immeasurable vastness may assail him like the wind—a wind perceived for the first time as a mighty elemental surge. Entranced by the gigantic and eternal drama that is the sea, he will give himself up to the monotonous roar of the waves and the shout of heavy waters battering the shattered cliffs ... to the sweeping curve of the bay, the rocks like the bodies of animals, the leaden reflections in the faded waters of the horizon.

Saying he needs fresh sea air, he will return once more to the sea, this time with his family, the children armed with the touching toys of childhood: sailing boats, buckets and spades. Bravely, he will throw himself into the waves, feel the ground disappearing beneath his feet, and emerge spluttering from the spray. Rejuvenated, or so he feels, he will throw

New Orleans around 1873

himself once more into the water, groping in the glassy wall of the billows, invigorated and lulled by the support of the sea.

Later he will discover the ocher beaches of southern shores lying at the foot of sun-warmed rocks, and the tender crescent of a hidden bay where fishing boats doze at noon ... the cool sparkle of the morning sky, and the last tremble of the setting sun ... lagoons bathed in light, raging whirlpools among Cyclopic rocks, the caustic blow of salt upon the skin, and the almost tender caress of the sun.

He will experience the whisper of grass in the dunes, the singing of the wind among the pines, and the flutter of the breeze in silken pennants ... the iodine scent of seaweed, the delicious sharpness of scattered salt, and the unexpected thrust of a sudden squall before the breakwater.

The urge may come to him to venture out beyond the green shallows onto the heavy, breathing swell of the open sea, on toward the immeasurable space in which abyss and sky are separated only by a constantly retreating horizon. And perhaps he will experience a storm—a storm that approaches from a black sky, bringing gusts of wind that batter like fists on the tarpaulin ... and a rush of menacing waves, the crash of the bare bow of the boat on the black water, flashes of chalky lightning—and fear. Thus he will discover the power and vastness of the sea, an element of creation that conceals so many mysteries and marvels in its unplumbed depths that its contemplation overwhelms us with wonder and a long-forgotten awe. Awe—a word that embraces so much that is deeply felt: the humble veneration of the divine and the recognition of an unearned gift, a gift that will endure forever: the timeless magic of the everlasting sea....

TROPICAL COASTS

Never before had I experienced such a feeling of confinement as I did on this sea voyage from Genoa to Tanga. It was like life in a hotel when the guests begin to know each other rather too well, and such details as the humble and reproachful grace that the American missionaries at the next table said before meals and the flirtations and intrigues at the dances in the second-class salon had become irritating. And all exits led to the water. The confusion of the streets and bazaars of Port Said was long since past; the conversations on the bulwarks during sleepless nights on the Red Sea forgotten; the Europe to which we had once belonged had sunk beyond recall, submerged in the unceasing flood of the wake and in the hasty sunsets. All of us, from the young emigrants to the "old Africans," all of us were eager for our destination, for the coast, harbor, land, green vegetation and odors of the earth. But as our steamer approached the velvet surf of the Indian Ocean and set her course toward the south, it seemed as if the land had abandoned us entirely, and day after day the *Usambra* steamed on in solitude under a radiant white sky, through a monotonous, endless circle. Even the untiring dolphins had fallen behind at Bab el Mandeb. And now our ship was no more than a bow plowing into the water, forcing the waves apart and burying them behind us in clouds of foam.

On the third or fourth day, the coast should have been visible; all we could see, however, was an impenetrable bank of mist hanging over the water like a wall, chalky, gray and cold. Thrust seaward by the wind, it tumbled over the decks and funnels, tangled in the derricks and the wheelhouse, settled firmly and densely on the shuddering body of the steamer, and smothered the last gleam of day in rain-glistening dark. The sirens signaled ineffectually through the abyss as our steamer circled restlessly hour after hour in the cheerless darkness. Then, at last, the pilot boat that had sought us untiringly appeared suddenly out of the gloom, its headlamps piercing the blackness like yellow-nosed rockets.

THE PICTURES

As our anchor finally sank in the bay, the last rain showers sped over the agitated water and past the throng of curious, excited fishing boats and cutters, and the gleaming pennants of the palms were revealed waving gently in the late afternoon light from the hills.

Anyone who has lived for any length of time in close proximity to the sea, particularly along a tropical coast, will never forget his first impression of it. Intertwined in the boughs of strangely shaped trees for which he knows no name, the huge expanse of the sea lies spread out in the distance, far away yet inexplicably close. The wind shakes the branches of a lonely Borassus palm on the edge of the sisal plantation, rattling them like wooden shutters; it is the trade-wind, which blows in from the sea and sweeps over the coast and hinterland hour after hour, morning and afternoon. All the winds come from the sea; and from it too comes the white light of the sun when the clock strikes six, abruptly transforming the night with a flood of breathtaking heat.

There lies the sea; and for the white people who are used to it, it has little importance apart from its usefulness in bringing ships to their harbors and providing transport for the things that the Negroes wrest from the earth—copra and palm oil, coffee and peanuts, and sisal pressed into huge bales and loaded onto lighters when the tide is high. But the lust for adventure always drives the newcomer into the jungle of mangroves, which is neither land nor sea and which has no ground and no sky—a knotted, matted wilderness where crocodiles, yellow-eyed and invisible, follow his log canoe and the gleaming bush knife of the Negro in the bow sweeps the sleeping tree snakes from the hanging roots. And when the light of the full moon drips from the black-green leaves of the mango trees and the sweet poison of luxuriant, night-scented flowers numbs and intoxicates his senses, he flees to some hidden, sandy bay where the night wind sighs and whispers in the palms; and there, beneath the moon, flings himself naked into the incomparable joy of the waves and gives himself up entirely to—the sea!

OFFSHORE

The bay was anchor, sky
and island: a land's end
sail, and the world tidal,
that day of blue and boat.

The island swam in the wind
all noon, a seal until
the sun furled down. Orion
loomed, that night, from unfathomed

tides; the flooding sky
was Baltic with thick stars.
On watch for whatever catch,
we coursed that open sea

as if by stars sailed off
the chart; we crewed with Arc-
turus, Vega, Polaris,
tacking into the dark.

Philip Booth

STORM AND STILLNESS

THE PICTURES

Speculation as to the exact moment in history when the first sail was hoisted is as idle as the question of when it was that man first traveled across water in a boat. We know from the evidence of cave drawings that rowing boats existed as early as the early Ice Age, but when did man first erect a mast and hoist a sail? And did sails exist before the invention of the loom—sails made of plaited bast, grass, or rushes? Streamlined Egyptian sailing ships, additionally equipped with oars, sailed the Mediterranean five thousand years ago. And the outriggers of the inhabitants of the South Sea Islands, the pre-Columbian Indians' drop-keeled sailing rafts, and the Chinese junks—at what period in history did they conquer the sea? Perhaps it was the wind itself that first gave man the idea of the sail; when he felt its impact and energy on his outstretched hand he may, in a moment of deep perception, have visualized its power harnessed to a matting of rushes or papyrus or, in other words, a cloth.

Sailing vessels conquered the world, and as long as men exist who are strong enough to rig a boat, hoist a sail, and hold a tiller, white pinions will continue to propel soughing boats over the water, on rivers and lakes and on the sea. The age-old dream of going to sea as a ship's captain, a conqueror, a discoverer, a pirate of time and space and lord of the wind and storms continues to exist, and the indomitable lust for adventure is still alive. There are many who have tried it alone and fought against storms, hunger, thirst, and loneliness, sailing over immeasurable oceans beneath a merciless sun, gallants, and die-hards: Henry de Montfried, Eric de Bisshop, Alain Gerbault, René Lescombe, Sir Francis Chichester, and many nameless ones who perished. Training ships—first-class three-masted barques, similar to their sisters of the last century, where the ancient craft of the mariner is taught in a hard school, still exist today, ships where discipline is of primary importance and the future ship's officer learns how to react in an emergency when

his ship runs into trouble. But surely our modern steamers are invulnerable? In his sea story entitled *In Hazard,* Richard Hughes describes what can happen to a modern ship that is caught and lamed by a hurricane and driven over shallow waters and coral reefs for seven days:

"The air was gaspingly thin, as on a mountain: but not enlivening: on the contrary, it was damp and depressing; and almost unbearably hot, even to engineers. Big drops of sweat, unable in that humid air to evaporate, ran warm and salt across their lips. ...

"For the first time, since the storm reached its height, they could see the ship from one end to the other. For the first time they saw the gaping crater left by the funnel's roots. Smashed derricks, knotted stays. The wheelhouse, like a smashed conservatory. The list, too, of the ship: that had been at first a thing felt: then, as they grew accustomed to it, almost a thing forgotten; but now you could see the horizon tilted sideways, the whole ocean tipped up at a steep slope as if about to pour over the edge of the world: so steep that it seemed to tower over the lee bulwarks. It was full of sharks, too, which looked at you on your own level—or almost, it seemed, from above you. It looked as if any moment they might slide down the steep green water and land on the deck right on top of you. They were plainly waiting for something: and waiting with great impatience.

"But the sharks were not the only living things. The whole ruin of the deck and upper-structures was covered with living things. Living, but not moving. Birds, and even butterflies and big flying grasshoppers. The tormented black sky was one incessant flicker of lightning, and from every mast-head and derrick-point streamed a bright discharge, like electric hair; but large black birds sat right amongst it, unmoving. High up, three john-crows sat on the standard compass. A big bird like a crane, looking as if its wings were too big for it when folded up, sat on a life-boat, staring through them moonily. Some herons even tried to settle on the lee bulwarks, that were mostly awash; and were picked like fruit by the sharks. And birds like swallows: massed as if for migration. They were massed like that on every stay and hand-rail. But not for migration. As you gripped a hand-rail to steady yourself they never moved; you had to brush them off, when they just fell.

"The decks were covered in a black and sticky oil, that had belched out of the funnel. Birds were stuck in it, like flies on a flypaper. The officers were barefoot, and as they walked they kept stepping on live

It keeps eternal whispering around
Desolate shores, and with its mighty swell
Gluts twice ten thousand caverns, till the spell
Of Hecate leaves them their old shadowy sound.
Often' tis in such gentle temper found,
That scarcely will the very smallest shell
Be moved for days from whence it sometime fell,
When last the winds of heaven were unbound.

John Keats

birds—they could not help it. I don't want to dwell on this, but I must tell you what things were like, and be done with it. You would feel the delicate skeleton scrunch under your feet: but you could not help it, and the gummed feathers hardly even fluttered.

"No bird, even crushed, or half-crushed, cried.

"Respite? This calm was a more unnerving thing even than the storm. More birds were coming every minute. Big birds, of the heron type, arrived in such numbers, that Captain Edwardes, in his mind's eye (now growing half delirious), imagined the additional weight on the superstructure actually increasing the list: them arriving in countless crowds, and settling, and at length with the leverage of their innumerable weights turning the *Archimedes* right over, and everybody sliding down the slippery decks to the impatient sharks. Little birds—some of them humming-birds—kept settling on the captain's head and shoulders and outstretched arm, would not be shaken off, their wings buzzing, clinging with their little pin-like toes even to his ears. . . .

"When at last the blast came, from an opposite quadrant, sweeping all those birds away to destruction, everybody was heartily thankful. Thank God not one of them was ever seen again."

Ships will always be vulnerable, like everything else on land or water, like everything that exists in the storm of time, for nothing on earth endures forever. But perhaps, did we but know it, ageless ancient gods are still enthroned in the glassy stillness of the deep! Poets perceive the symbolic power of the ancient divinities, the immortals, which is one with the element of water and its creatures and which preserves its myths for us in secret silence. . . .

Distress at sea

"Five times since we had dipped our bending oars
beyond the world, the light beneath the moon
had waxed and waned, when dead upon our course

we sighted, dark in space, a peak so tall
I doubted any man had seen the like.
Our cheers were hardly sounded when a squall

broke hard upon our bow from the new land:
three times it sucked the ship and the sea about
as it pleased Another to order and command.

At the fourth, the poop rose and the bow went down
till the sea closed over us and the light was gone."

DANTE ALIGHIERI
(translated by John Ciardi)

THE ARGONAUTS OF THE WEST

Our knowledge of the past is largely acquired through the study of documents that have been handed down to us and that spotlight personalities and events in the darkness of time. Among the most valuable and exciting eyewitness accounts at our disposal are the travel descriptions of Marco Polo (1254–1324); the one existing transcript of the logbook of Christopher Columbus's first voyage, made by his fellow traveler Las Casas, the militant "Apostle to the Indies"; Columbus's letters to the King of Spain in which he describes the success of his third voyage and endeavors to depict it in the most advantageous light; and the sketches of Antonio Pigafetta, the knight who accompanied Magellan on his first voyage round the world. The ingratitude and ill-will of the political rulers and the harsh fate met by many of the "argonauts of the Renaissance" is typified by the example of the Venetian traveler Marco Polo, who was obliged to dictate his experiences in Asia in 1298 to his cell companion in a Genoese prison! And Columbus, once admiral of a proud fleet of caravels and discoverer of a new continent, died in poverty and neglect.

When we maintain that some navigators of the time were luckier than others, we mean that in their seafaring ventures they were more favored in their political connections and by having personal fortunes. Although we know nothing definite about the appearance and constitution of the Venetian traveler Luigi Ca Da Mosto (born in 1432?), we may assume that he was a promising young man of good upbringing, education, and means who, according to his own words, was young and strong enough to bear the privations of a voyage of discovery along the Atlantic coast of West Africa. On a voyage to Flanders, he broke his journey in Portugal in order to meet the infante Henry the Navigator, known as an enthusiastic patron of explorers, cartographers, and argonauts, and to seek support for his projects. The infante provided him with a caravel of around ninety *bottes*, ships of approximately forty-five-ton carrying

THE PICTURES

capacity. The cost of the merchandise that Ca Da Mosto carried and with which he hoped to engage in exchange trade was to be borne by the twenty-two-year-old Venetian himself. Henry the Navigator was a businessman to be taken seriously, who was experienced in calculating percentages. But Ca Da Mosto, quite apart from his love of adventure and discovery, was also well aware of the financial aspects of the expedition, and he wrote in his logbook that he hoped to gain both honor and profit from the undertaking. On March 22, 1455, he sailed from Cape St. Vincent with a favorable wind and "in the name of God and full of hope" toward his first destination, Madeira, which had been in Portuguese hands for twenty-four years.

Inward and outward to northward
and southward the beach-lines
linger and curl
As a silver-wrought garment
clings to and follows the firm
sweet limbs of a girl.
Vanishing, swerving, evermore
curving again into sight,
Softly the sand-beach wavers away
to a dim gray looping of light.

Sidney Lanier

To the south was the great unknown, the seductive and delicious danger of adventure. And profit. And somewhere lay the legendary kingdom of the archpriest John of Abyssinia. Gradually the world became a sphere.... Or was it still heresy to believe that one could travel round the world like a finger round a ball, and that he who set sail to the west would land one day in India, or on the Moluccas with their herbs and gold and jewels, precious wood and slaves?

Ca Da Mosto wrote that the island of Argin had been let for the past ten years to Christians who owned not only settlements on the island but also trading posts for transactions with the Arabs from the mainland coast. The Christians dealt in woollen goods, cotton, silver, carpets, blankets, and maize—the Arabs were always hungry! In return they received gold dust and slaves brought from the Negro countries. Thus it was that seven to eight hundred slaves were shipped from Argin to Portugal each year—not necessarily because Ca Da Mosto was particularly interested in slave trading, but because dealing in Negroes was a customary aspect of expeditions of this kind.

After he crossed the mouth of the Senegal, which separated the land of the "gray-skinned Berbers" from that of the Negroes, and reached the southern land of Budomel, or Vedamel, he made the acquaintance of the Negro ruler—and brought off an excellent business deal. In exchange for horses and other items that took the king's fancy, he received not only a hundred slaves but also a very young and beautiful Negress. According to Ca Da Mosto, he undertook his visit to the king's residence inland not only for commercial reasons but also for the purpose of finding out everything worth knowing about the country and its people. After he had had a chance to look around and conclude his business, he decided to continue on to Cape Verde to discover fresh horizons and try his luck on new shores. His interest lay principally in the gold that, according to the testimony of Negro slaves, was to be found in the southern land of Gambra, or Gambia. On the way, he met up with the Genoese nobleman Antoniotto Usodimare, who commanded two caravels and was evidently also intent on discovery, gold, and slaves. Continuing the voyage together, sailing steadily with the wind along the coast, they discovered a river that Ca Da Mosto named Rio di Barbazini, and finally reached the land of Gambia. The arrival was not without its difficulties, however: one of the Negro interpreters with them took it into his head to start killing off the natives, and when they advanced farther into the mouth of the river Gambia they encountered Negroes in log canoes who not only refused to trade with them but also shot at them with poisoned arrows. When the white men replied with volleys of cannon balls and arrows relations were hardly improved, and the situation was further worsened by the fact that the natives had heard of the arrival and the business methods of the white men from the Negroes of Senegal. The two navigators deliberated whether or not to continue up the river and penetrate into the interior of the country in the hope of finding more reasonable inhabitants, but the sailors had had enough of hazards and adventure, and the flotilla set course towards Cape Verde and the harbors of home.

Ca Da Mosto returned to Venice in 1463, where he died in 1483 or 1511. The account of his voyage has been handed down to us in the *Navegaceões de Luigi de Cadamosto* (published in Lisbon in 1812), a valuable document of the age of discovery. There is no doubt the Ca Da Mosto was more than just an adventure-loving seafarer; his travel book contains much worthwhile ethnical information about the Atlantic coast of Africa, and he also had an eye for the beauties of nature—even if he himself did not value this aspect of his voyage as highly as the financial advantages he gained through his encounter with the king of Budomel.

But was Ca Da Mosto the first to discover and explore this coast?

According to Greek tradition, the Carthaginian navigator and colonizer Hanno wrote a report on a voyage he made to the shores west of the Pillars of Hercules in the fifth century B.C. This expedition of sixty five-oar ships was undertaken with the purpose of founding off-shoot towns and commercial settlements. Thus the Carthaginians pursued their policy of expansion as successors to the Phoenicians, the "conquerors of the sea." The undertaking—gigantic even by contemporary standards—was carried out by approximately thirty thousand persons and was inspired solely by a thirst for conquest and power; but as far as we can gather from Hanno's report, the advance into this uncharted territory, fraught as it was with the unknown dangers of crocodiles, sea monsters, drum-beating savages, hairy apes, and volcanoes with glowing rivers of lava, must have been terrifying in the extreme. These argonauts reached the island of Fernando Poo in the Gulf of Guinea and probably only turned back due to lack of food.

Sailing boats in harbor
(after Chodowiecki)

An unknown ocean had been navigated, a wide new horizon discovered and for a time possessed, and man would not rest until he had sailed round the whole world and the West had become the East.

Much has been written about the tragic fate of Christopher Columbus—tragic not because he did not succeed in reaching India from the west, but because of the Spanish rulers' unscrupulous exploitation of his courage and indomitable spirit for purely commercial and economic reasons. They were interested not in the confirmation and expansion of geographical knowledge but in the acquisition of rich, gold-yielding colonies. The fleet of seventeen ships and the small army with which Spain provided Columbus for his second voyage was not intended to serve as a geographical expedition: its job was to pursue a Portuguese fleet headed for Haiti. Spain and Portugal were wrestling for world power, and the sea was the decisive element. Anyone who was willing and strong and brave enough to take command of a ship and a troop of rough, undisciplined soldiers had troops and weapons put at his disposal. He might win respect and fame, money, honor, and a title, or he might perish in nameless oblivion. But as long as there were captains who desired nothing more than the movement of the decks beneath their feet, kings, counts, doges, and statesmen had no need to fear for their power.

The enormous deployment of power kept pace with a development of knowledge that was unprecedented; in the very same year that the roar of the cannons of Columbus's *Santa Maria* shocked a continent out of its slumber off Haiti and Cuba, Leonardo da Vinci sat in the silence of his lonely chamber working on the sketches of a flying machine.

The world was expanding, and there seemed to be no spatial boundaries and no limits to the potentialities of the mind. The Tartar count and astronomer Ulug-Beg built an observatory in Samarkand in the fifteenth century and equipped it with huge instruments to enable him to determine the exact positions of over a thousand stars of the Ptolemaic system. During the next two centuries, warlike disputes and battles for power between East and West and North and South followed one after the other. In 1512 a rapidly evolving England built ships of a thousand tons equipped with seventy guns, English pirate ships captured over four hundred French ships in the Channel alone, and Italian and Spanish fleets destroyed the Turks' sea power. The Spanish Armada succumbed

Launching of a ship

THE PICTURES

Page 43	Figurehead of the *Cutty Sark*
Pages 44/45	International training ships, Kiel Week
Page 46	Star boats heading into the wind

to the English fleet under Sir Francis Drake and lost one hundred and sixty ships to the value of sixty million ducats; and the foundation of English, Dutch, and French East Indian trading companies as real limited-partnership companies with trading monopolies that yielded twenty and twenty-five per cent interest ensued in the same year.

And what of the New World in North America?

Was it really the Vikings who fought their way through icy storms and drift-ice from the mysterious North to the far West? Was it a *drakkar,* with alarming dragons on the bow, that first ran aground on the North American coast? We can read of the great navigator Eric the Red, who sailed to Greenland in 985 A.D. with twenty-five ships and a cargo of men, women, cattle, wood, and hay; and tradition has it that his son Leif set out in a westerly direction in the year 1000 with a few companions and discovered a land where there was no frost in winter,

CUTTY SARK

SR
4773
5095
5098

where the days and nights were different from those of Greenland, where the summer was warmer and the pastures green and luxuriant. And where wine grew! Thus the Vikings called the new land "Wineland"; but when they returned some time later, they were met by an unknown, hostile people, the Red Indians ... and so the story of their discovery was forgotten for centuries.

Behold the threaden sails,
Borne with the invisible and
creeping wind,
Draw the huge bottoms through
the furrow'd sea,
Breasting the lofty surge.

William Shakespeare

It was a different race of people, with other ships and weapons more effective than the Vikings' swords, that finally succeeded in holding its own in the West. From Sir Walter Raleigh's first settlement on American soil, the first English colony arose in 1607: a great power that was to rule the seas for centuries was on the brink of conquering a new world.

But it took more than navigators and discoverers to create something permanent and valuable out of the New World. The conquerors were followed by pioneers and colonizers who urbanized the land and grew with it, and the evolution of the new state was dependent upon a total spiritual, mental, and geographical penetration of the North American continent. From the mixture of races that flocked to the new continent, in the crucible of a spiritual, economic, and military testing ground, a new nation was born: America!

When, after approximately three centuries of attempts to reach the West, the Pacific Coast was finally attained, a new horizon was opened up to this new nation. The Pacific Ocean was eventually to induce the young state to more and more power. But the advance toward the new horizon led at first to a confrontation with the very power that also ruled the East: ultimately, inevitably, the last word lay with the sea.

HARBORS LIKE HUGE NETS

In harbors all over the world there are—and always have been—people who belong to no ship's crew; who have nothing to do with the loading and stowing of crates, barrels, and bales; who are part of nothing and belong nowhere in this jungle of noise and smoke: the emigrants. In cold, dismal halls, they wait for the hour of farewell; mahogany-framed pictures on the wall show ocean liners steaming proudly out to an azure sea: fine gentlemen in white suits lean over the railing, and delicate green veils lie softly on the cheeks of beautiful women. ... But the emigrants do not sail in ships like these. Their ship looks as dilapidated as the passengers who now hesitantly board her, pushing small suitcases and bundles bursting at the seams before them with their knees, up onto the deck where an officer takes their papers without a word before the steward leads them down again into the steerage, as into a tomb. At the first blast of the ship's siren, they ascend to the deck once more to cast a last long glance at the town. Slowly the ship frees herself from the quay, invincible and relentless. And the hearts of the young ones among them throb with the rhythm of the ship, infected by the pulse of the engines, and they forget the dull ache of farewell in a new intoxication as they move toward the open sea and an unknown destination.

Harbors are always the scene of arrival or farewell. They are like nets that ensnare men in the river of time, retain them for a while, and set them free again; reloading points where cargoes of men and merchandise are loaded and shipped. And always they are fascinating; always we are lost in wonder at the sight of such a giant toy, a toy composed of steel monsters, full-bellied hulls, streamlined launches, rusty tramp steamers, broad-breasted tugs, and a tangled web of ropes and cables over the brown, splashing, oil-stained water.

I remember Massaua on the Red Sea at the hour of the glowing evening sky. There were giant cranes like giraffes leaning thirstily over the

THE PICTURES

water, hoarse calls of traders on the quay, the nasal cry of a camel, and, in a sudden silence, the hot breath of the earth; and the odor of eating places, fermenting bananas, and smoked fish.

Then there was Aden, gray with the desert dust of centuries. Gray and shimmering with heat were the walls and docks, the rocks sheltering casemates and arsenals in their hollow interiors. And the huge open cisterns, descending in giant steps to house-high depths, empty except for a brackish puddle or two. ... Alongside the pier lay a cruiser, perhaps the *Enterprise*. She had come from Durban or Hong Kong or Malta and was refueling now with oil brought to her in gulps by thick, snakelike tubes from gray tanks. Aden was a halfway house, a barbed-wired milestone on the sea routes of the British Empire, and the world's biggest coal bunker, in days gone by....

Old engraving of the harbor of Zara (now Rieka, Yugoslavia)

In La Rochelle a wind freshly washed by a night shower sped merrily over the squares, tugging at the headscarves of the market women and robbing the old men of breath. Everything was clean and shining, bright and healthy again after the war. On the Ile de Ré, salty white drops of water fell from the crests of the waves like cast seed.

And Belfast. A gentleman from the Harland and Wolff wharf administration building asked us for our cameras, and so we were obliged to register our impressions by eye and ear alone: the human ants in the groaning depths of a ship's hull like an upside-down dome of iron ... the harsh stuttering of a riveting hammer and the almost churchlike stillness of the workshops where, next to a milk bottle and teapot, a cat sat blinking on every work bench. ... Near the docks a troopship, farewells and weeping soldiers' brides, and an old man offering everyone a drink from his bottle.

Arriving at Portofino was like entering the auditorium of a theater in which a performance is due to begin any minute—Goldoni, perhaps, or *The Bride of Messina*. The "chorus" consisted of fishermen mending their nets, handsome sailors, market women crocheting lace, wandering carpet-traders and short-skirted girls, ready and willing. The backdrop was colorful, freshly painted rows of houses and yachts overlaid with a golden sheen in the enchanted blue of the mild evening. And the air was fragrant with flowering bay trees and sun-warmed thyme....

THE LURE OF THE DEEP

Somewhere at the bottom of the sea lies the legendary land of Atlantis and, as if it were a key to an understanding of their own existence and significance, men never tire of seeking this sunken island. In the nocturnal darkness of the deep, reposing in the corroded hulls of wrecked ships and iron-bound chests, lies a fifth of all the gold and silver ever mined by man. In 1682 one man alone succeeded in recovering two hundred *livres d'or et d'argent* from a Spanish galleon, and in 1927 the crew of the salvage ship *Artiglio* reclaimed hundreds of tons of copper, steel bars, whole railway trains, and seven locomotives from the sunken wreck, close to a precipice, of the *Washington,* from depths to which only helmeted divers or diving bells dared venture.

But long before that, men from classical antiquity had returned from the purple vaults of the sea with tales of strange creatures and plants, breathing sponges, iridescent flowery fans like spun glass, shoals of rainbow-colored fish, menacing sharks; nevertheless, they brought to the surface oysters with pearls embedded in their pink flesh. The pale luster of the pearls, strung into flattering necklaces, graced the skin of queens and courtesans; and delicate links of coral, cleverly knotted, lay blood-red on the breasts of nameless women of fisherfolk throughout the world.

Men of our century, too, experience the consuming desire to venture into the blue-green depths, to dive deeper and deeper and to glide almost weightlessly through a strange and unknown world that is like nothing known on land. It is a world where strange plant-animals grow from the rocks; where transparent, glassy jellyfish move weightlessly through the noiseless void; where azure fishes flit over structures as nameless now as on the first day of creation—sulphur-yellow and red, turquoise and star-blue—and where delicious danger and new, unknown adventure lurk.

Man has discovered a new world beneath the sea, has encountered the unfathomable, the unimagined. But when he has tried to bring these treasures of pure color and fantastic form to the surface and expose them to the light of day, he has been forced to realize that they lose their splendor in the hard light of the sun and, robbed of life, disintegrate. Many a diver has deserted his harpoon at the approach of a shark, or when the intoxication of the deep made him stagger and fall. But no one who has ever experienced it can forget this world beneath the sea, and he will return again, time after time. Why? To feel like the fish he may have been millions of years ago? To become one with the water, the element from which all life originates? Or for the sake of a rare magic, a mysterious enchantment?

Alexander the Great tries his hand at diving

There is something of the discoverer and explorer in all men, and many of those who do not aspire to acquire wings and imitate the birds bind flippers to their feet and artificial lungs to their backs and become for a while protégés of Poseidon or sisters of the green-eyed sea nymphs. Many divers have learned to take cameras with them instead of harpoons, and they bring back from the depths images of an enchanted world. Shoals of gleaming blue coral perch glide over pastel coral reefs, barracudas jerk like iron-colored lightning through the green twilight of ocean lagoons, and starfish gleam purple in the bluish night; the parrot fish stares in surprise as he casually crushes tiny fragments of coral in his beaklike mouth, and a sea hedgehog stares motionless at the camera, aware of his invulnerability; a prickly globe fish is huge and alarming in his blown-up state, and confident in the knowledge that he can kill. Like poisonous flowers, mallow-colored sea anemones entice small, unsuspecting fish into their deadly mouths; and when the blue shark rises from the deep and roams ravenously through the coral gardens, everything that has fin or foot takes shelter, flees, or hides, and man—like the others an enemy and intruder—reaches for his dagger. For danger approaches from the deep: the shark is a pirate roaming restlessly, mercilessly, through the fertile waters of the coast. At this point I am reminded of the story of the somewhat naïve Englishman who, after swimming unconcernedly in the blue waters of the Indian Ocean, expressed his surprise at having seen no crocodiles. On learning that, whereas crocodiles never leave their rivers, imposing numbers of sharks cruised regularly near the beach, he very nearly fainted.

Not all the stories of giant octopuses with eight tentacles armed with suckers attacking men and boats can be discounted as medieval rumors or seamen's yarns. In October 1873 a family fishing in a dory between Bell Island and Portugal Cave was attacked by an enormous monster that clasped the helpless little boat in its huge tentacles and very nearly.... Fortunately, however, one of the more stouthearted members of the party saved the situation by hacking at the octopus's tentacles with an ax before the boat was crushed. One of these tentacles, around nineteen feet in length, is preserved in the British Museum in London. The coast of Newfoundland seems to be a favorite playground for octopuses, although they—or just their tentacles—are usually found washed up dead on the shore. One such monster is said to have measured thirty feet in length and to have weighed three hundred and thirty-one pounds. In 1966 a gigantic battle took place between a giant octopus and a whale, the outcome of which is unfortunately not reported.

The sea is ageless. How nicely that reads, and how poetic it sounds! But in fact the sea is subject to the tides of time just as mountains and continents are. Probably only the element of water itself is truly ageless, because it is constantly rejuvenated and renewed. Since prehistoric times, the winds have driven clouds born of the sea to the land; rain and snow fall to earth, disperse, and flow back to the universal mother, the sea. And for longer than we can possibly imagine, giant currents have been moving over the floor of the sea, flowing between gigantic mountain ranges that lie beneath the water in eternal night—a precipitous underworld from which volcanoes occasionally erupt. Compared to these currents, our continents are mere islands, the wreckage of millions of years. And above the currents flow the waters that we have given names and that unceasingly wash the shores of land, of faraway Greenland, Alaska, Asia, Antarctica; waters influenced by the rotation of the earth, the sun, the winds that sweep the world, and the mysterious phases of the moon. Even creatures born of and nourished by the sea are subject to the currents and wander, get lost, die, and fall to the deep. Seas that once existed have disappeared and been forgotten even by history, and only someone who knows how to interpret the markings on fossils, to tell the age of volcanic basalt and analyze sediments after thousands of years can reawaken their memory. Many of the discoveries from primeval times affect us almost like miracles. A lake in southwest Yugoslavia, once part of a huge sea that stretched from the Atlantic

THE PICTURES

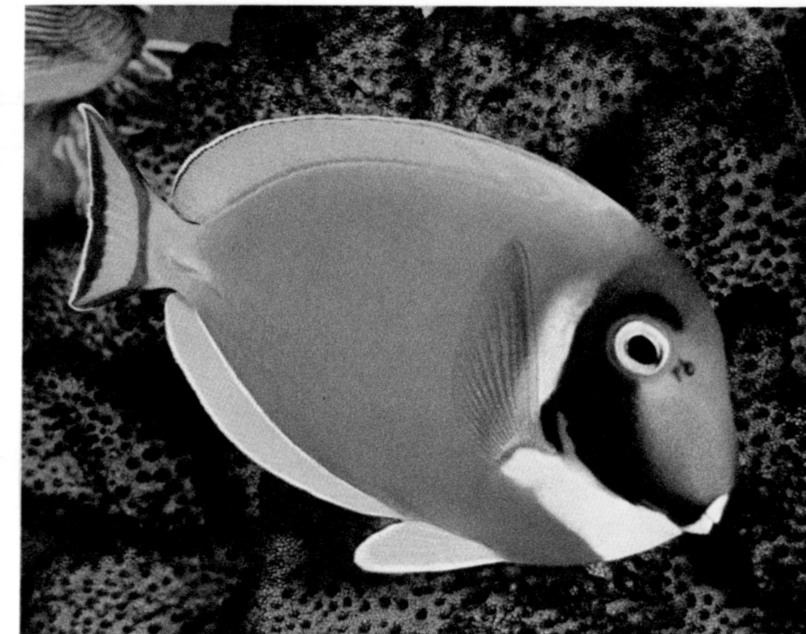

THE FISH

Wade
through black jade.
Of the crow-blue mussel shells,
one keeps
adjusting the ash heaps;
opening and shutting itself like

an
injured fan.
The barnacles which encrust the
side
of the wave, cannot hide
there for the submerged shafts
of the

sun,
split like spun
glass, move themselves with spot-
light swiftness
into the crevices—
in and out, illuminating

the
turquoise sea
of bodies. The water drives a
wedge
of iron through the iron edge
of the cliff; whereupon the stars,

pink
rice-grains, ink-
bespattered jellyfish, crabs like
green
lilies, and submarine
toadstools, slide each on the other....

Marianne Moore

between Europe and Africa over the Near East to the Himalayas is, although its waters are no longer salty, still the home of species of fauna that lived over twenty million years ago; and in the Viennese Basin, which thirty million years ago was covered by the sea, scientists have discovered a huge fossilized pearl.

The sea keeps its secrets. How true, yet how erroneous! Islands, great stretches of land, towns and ships have sunk in the sea, and over them lies a watery silence and a great forgetfulness. Or at least until a diver returns from the deep and tells of centuries-old objects that he has found: fallen columns, fragments of a frieze encrusted with mussels, or a statuette; or maybe even a sealed amphora, redolent of Attic wine, bound for a foreign shore long ago.

As children, our greatest treasure was a snail-shaped, iridescent shell, precious because when we held its mouth against our ears we could hear the roar of the sea. No wonder, really, for it used to belong to an uncle overseas who left it to us with a final greeting. Shells like this are treasured by many who have long since lost their childish faith. The Tritons of Greek legends used to blow on conch shells, and shell bowls were used as drinking vessels and the basins of fountains. And when John the Baptist used a shell to sprinkle the waters of baptism over the faithful, the sound of the rushing of eternal waters was preserved in them forever.

THE MIRACLE OF CREATION

The miracle of creation is the miracle of an all-embracing, all-sustaining law and order, and the deeper science delves into this "divine plan," the more miraculous it seems and the more apparent it becomes that everything has a meaning, even the evanescence of human life. Obscure though the words of Genesis may sometimes seem, the story of the creation is a clear description of the birth of the world as a planned and ordered development, as birth and growth. No matter how each individual interprets the actual words, no one can fail to grasp their deeper meaning: "Darkness was upon the face of the deep. And the Spirit of God moved upon the face of the waters"—clearly a declaration that the Spirit existed from the beginning. And there was light, and day, and a firmament in the midst of the waters, the heaven. "Let the waters under the heaven be gathered together unto one place, and let the dry land appear." And as the sea was created, God said, "Let the waters bring forth abundantly the moving creature that hath life, and fowl that may fly above the earth in the open firmament of heaven. And God created great whales, and every living creature that moveth, which the waters brought forth abundantly."

Science has investigated this abundance, and in the last four hundred and fifty years no less than three hundred thousand species have been classified, twenty thousand of which are bony fish. And the sea supports them all—or rather its fertility is so great that vast amounts of superfluity and waste pass unnoticed. For example, the cod lays five hundred eggs, the oyster five hundred million, and the progeny of certain algae can reach billions in a single month. Why? Simply so that they can survive in the mighty and relentless cycle of "Who eats whom?" Nature is always extravagant, and her creatures live in a continual process of procreation and development in which each is preyed upon by its neighbor. A well-fed sperm whale may have as many as five thousand small herrings in its stomach, each of which may have swallowed six or seven thousand tiny crabs, which in their turn have eaten over a hundred thousand algae. Man, for whom two dozen oysters or a lobster is an ample meal, lives modestly by comparison. But the number of people in the world also runs into billions, and each individual needs to eat, if not as extravagantly as the whale. The sea is inexhaustible, or so we are told. It is only a question of making better use of it. One day we will be able to produce human food from sea plants. One day ... One day even the whales may become extinct, although it is true that whalemen reach agreements from time to time to catch only so many blue whales per season—not, I may add, for the sake of the whales but for purely material reasons. Unlimited whaling can lead only to whales becoming extinct, and although it is comforting to think that they can take refuge in the huge expanses and depths of the open sea, we must not forget that as early as the fifteenth and sixteenth centuries the Basques persecuted Greenland whales on the open sea and hunted them as far away as their home in the Arctic Ocean. And not only for their blubber, but for the fashionable whalebone and the sperm whale's gallstones, which were used for perfume.

Sea monsters (after Merian)

When the last Biblical sea monsters are seriously threatened with extinction, perhaps man will pause to reconsider his folly and hastily create reserves, as he has for big game and birds and sequoia trees and butterflies. But he will do so only when he becomes aware of the extent to which he has mistreated defenseless living creatures, or because he feels that something irretrievable is in danger. Or does man's concern about wildlife threatened with extinction have its origins in an ancient, atavistic sense of guilt that assails him when he kills? This does not apply to the fishermen who make their living from the sea, but to those of us who know, or at least suspect, that we cannot willfully destroy nature's order and balance just for the sake of thrills and trophies without paying the price—the price of a bitter sense of loss and a thoroughly justified fear that the animal world is becoming impoverished. In fact, it is only in the last century that we have learned how important wildlife is to us as a symbol of naturalness and of creation itself. To quote the words of a man who is still able to perceive nature as a priceless gift, "The aspect of greatness, exaltedness, that which is awe-inspiring or even frightening, has something of a healing influence on our spiritual arrogance and unthinking acceptance of nature."

The blessing of the sea is twofold: it is a great provider and it is also a theater, the scene of a drama played by the ever-changing sky, the endless expanse of water and the creatures which it cradles, the birds of the air, the light and the stars. And although fishermen and others who wrest their living from the sea may smile when we talk of a blessing and a gift, their smiles will be indulgent; for they are the last to be left cold by the mighty drama, and they live with it, participate in it, and know the fate and fortune of those bound to the sea. The rest of us stand by admiring, stirred and blessed by this vastness and abundance of life and by the magic of the sea, which we desire and seek like the enchantment of the mountains or the green wilderness.

But the sea's magic is different from that of the constantly rejuvenating woods or the mountains soaring to the sky. It has the fascination of the unknown, of secrets concealed in night-dark depths, and we feel that somewhere down there where we cannot reach there may still exist age-old beings from before the flood, primitive forms of life that need no light, prehistoric creatures born of night.

Not everyone has the good fortune to sail in Polynesian waters or to the paradisiacal Galàpagos islands in a yacht; nor is it easy to find a place on a whaler—especially as an onlooker—or to fly to Antarctica to take pictures. But flights to all tropical coasts and to Greenland and Iceland lie within the bounds of possibility for most of us, and I know of a man of mature years who completely lost his heart to Greenland because of the icebergs and seals that he hunted with his Leica in a kayak. And then there was the young photographer who spent a lonely summer on an island in the Hebrides and talked about the sea parrots and puffins much as one talks about well-loved relatives or friends; indeed, slightly malicious friends maintain that he is already fluent in "puffinish." ... Instead of walrus teeth and murderously toothed swordfish noses, stuffed birds and gulls' eggs, the skins of animals and the wings of birds, instead of sand crabs, sea urchins, and sea anemones that collect dust, dry up, and disintegrate—how much more worthwhile it is to bring back pictures of a sojourn with the sea, permanent records of a new experience, testimonies to a new way of thinking and feeling!

The dream of Moby Dick lives on, and the secret longing for danger and battle is never quite extinguished. Just as we can never quite forget our boyhood dreams of the *Odyssey* or *Robinson Crusoe,* we can never quite forget the desire for battle with the dragon, the sea monster, the

THE PICTURES

Page 81 Approaching storm
Page 82 Evening by the sea

whale in *Moby Dick*. But anyone who delves deeper and reads the description of an encounter with the sea monster written by the first mate of the whaler *Essex,* Owen Chase, cannot fail to experience a feeling of dread. Of the twenty survivors who managed to scramble into three lifeboats from the sinking *Essex,* only five reached land alive after drifting for ninety-three days; the rest perished, and many of those who died were eaten by their companions. When it is a matter of life or death—and Chase describes the distress of the shipwrecked crew with a terse and telling poignancy—nature will eat even her own children. It is interesting to note that it was once again an American who described a battle between a man and a giant of the sea: and after the tragic death of Ernest Hemingway it is easier to see the figure of the old man as a symbol of all lonely men in conflict with the powers of evil, the sinister forces of the deep that in the end destroy him.

To me the sea is a continual miracle,
The fishes that swim—the rocks—
the motion of the waves—the
ships with men in them,
What stranger miracles are there?

Walt Whitman

I have tried to write about the sea as the great provider and have ended wandering in daydreams: I, too, am a victim of the magic of the sea. And I nearly forgot that the sea is the provider not only of food, but also of energy and materials that man must one day harness if he is to satisfy the needs of all his kind: metals and ores, mineral oils and raw materials for the chemical industry, and enormous reserves of power in the mighty mechanism of the tides alone. And the original source of life-giving water—this too is the sea!

THE KEEM HARBOUR SHARK TRAP

And now I should like to tell a story.

It is a story about sharks in Ireland and, like all good Irish stories, it begins in a pub. In an Irish pub, nota bene, which is a mixture of a saloon, a cosy, musty waiting room, and a village shop. A tall, lean man wearing a leather jacket stood leaning against the bar, humming to himself. In spite of this, he did not look very happy, and he regarded the people around him with a disapproving expression. Particularly us—strangers who spoke with a foreign accent and stood around drinking only moderately. "If you were real reporters, you wouldn't be hanging about here," he said. "You'd be out at Achill Island where the sharks are. They catch sharks out there." He had heard that we were journalists and were traveling round Ireland from the landlord, who was filling huge glasses with beer.

"Sharks?"

"Yes, sir—sharks!"

"Man-eating sharks?"

He looked at us with contempt. Did we think that he, Jim So-and-so from Houston, Texas, U.S.A., had nothing better to do than wander around this God-forsaken coast and ask every shark he saw if it ate people? We assured him that we hadn't really expected that and thanked him for the tip.

In Achill Sound, at the end of a narrow bridge of land, stood Mr. Sweeney's General Store, where you could buy everything that an Irishman with money needed or, if he had no money, at least desired. For some time nobody mentioned the subject of sharks, although it appeared that Mr. Sweeney had a certain interest in them. The road to Keem Strand, where they caught the sharks, was not very good, we were told, but nevertheless passable with a certain amount of skill—we'd see for ourselves.

A sea serpent attacks the American two-masted vessel *Sally* off the coast of Long Island (December 17, 1819)

We did. For a whole hour we drove along a steep, slippery track on a rocky precipice above the sea. The sea itself looked something like a maelstrom gone mad. We tried to explain to two men who had just pulled a "curragh" onto the beach that we had come because of the sharks, and asked if they could somehow.... Somehow we suddenly found ourselves in a perilously rocking boat, headed for the bay.

We had tea below deck—very good, very hot tea, generously laced with rum. The skipper explained how he and his crew dealt with the sharks. "They come here every year—big fellows. They don't eat the fish or animals or crabs. They come in shoals from the south to the bay, and then they turn right round again and head for the north. At least, those we don't catch do. We don't know where they come from. We spread nets from the rocks to the bay. No, vertically! They run into the nets with their noses and then they rise. You only see their dorsal fins above the water. Then we kill them."

"With a harpoon?"

"With a kind of spear. You'll see."

I saw the gray, jerking triangle above the water first and cried, "There, there!" We slid down the side of the boat to the men who rowed with all their might toward the shark. When we were near enough, they pulled the oars into the boat and one of the men grasped the coarse mesh of the net with a sure hand, hoisted it up, and drew it to the boat. Then one of the other men thrust his spear into the spot where the head and neck of the shark must have been, three, four, seven times. The scarlet blood floated for a while like a pool on the bottle-green water. Only the liver was worth anything, only the oil. The liver of one shark had weighed a good twenty-two hundred pounds. The men from Keem Harbour—Needam, Grailes, Murphy, McGanty, and McConalty—shook hands with us as we left, and did not ask about the pictures in our cameras.

ACKNOWLEDGMENTS:

Lines from "The Last Voyage of Ulysses" from *The Divine Comedy* by Dante Alighieri, translated by John Ciardi.
Copyright 1954 by John Ciardi.
Reprinted by permission.

Lines from pp. 112–115 in *In Hazard* by Richard Hughes.
Copyright 1938 by Harper & Row, Publishers, Inc.
Reprinted by permission of the publishers.

Lines from "The Fish" by Marianne Moore.
Copyright 1935 by Marianne Moore, renewed © 1963 by Marianne Moore and T. S. Eliot.
Reprinted by permission of The Macmillan Company.

Lines from "The Marshes of Glynn" by Sidney Lanier reprinted by permission of Charles Scribner's Sons.

"Offshore" from *Margins* by Philip Booth.
Copyright © 1963 by Philip Booth.
Reprinted by permission of The Viking Press, Inc.

Below:
Galleys of the barbarians